Contents

CHAPTER 1
What are Monkeys and Apes? 4

CHAPTER 2
Lemurs 6

CHAPTER 3
Bushbabies, Lorises and Pottos 8

CHAPTER 4
Marmosets and Tamarins 11

CHAPTER 5
New-World Monkeys 12

CHAPTER 6
Old-World Monkeys 14

CHAPTER 7
Lesser Apes 16

CHAPTER 8
Great Apes 19

CHAPTER 9
Family Life 21

CHAPTER 10
Play and Grooming 24

CHAPTER 11
Life in the Forest 26

CHAPTER 12
Life on the Plains 28

CHAPTER 13
Danger to Monkeys and Apes 30

Glossary 32

Previous page: a mountain gorilla in the cloud forest of Virunga Volcanoes. Mountain gorillas are hairier than their lowland counterparts. Baboons (these pages) are extremely versatile primates, living equally well in humid rainforests, on rocky outcrops or, as here, in the bush.

1

What are Monkeys and Apes?

Monkeys and apes belong to a fascinating group of animals called primates. Zoologists organize animals into various groups according to how many things they have in common. For example, all animals which have fur, are warm blooded and feed their young on milk, are classed as mammals. The mammal class is then divided into smaller groups called orders, such as the carnivore order (which includes dogs, cats and bears), and the primates.

The primate order includes monkeys (nearly all of which have tails) and apes (none of which have tails), along with marmosets and tamarins, tarsiers, bushbabies, pottos, lorises and lemurs – in all a total of 183 *species*. All these kinds of animals are called non-human primates, because there is one other kind – the human primate.

What, then, are the things that make an animal a primate? All primates have both eyes facing forwards, so that the view seen by one eye overlaps that seen by the other. This "stereoscopic vision" gives a clear, three-dimensional image, and means that primates are good at judging distances – an important ability for life in the trees. Each eye looks out from, and is protected by, a round, bony case. Primates have hands instead of paws, often with a thumb or big toe which can grasp against the fingers or toes. With the exception of marmosets and tamarins, all primates have flat fingernails and toenails instead of claws. And compared to their body size, primates have a bigger brain than other mammals. Those with the biggest brains – monkeys, apes and humans – are known as "higher primates;" the others are known collectively as "lower primates."

The skeletons of many lower primates resemble the *fossils* of primates that lived about fifty million years ago. They are said to have "primitive features." As the higher primates evolved, sight, rather than smell, gradually became their most important sense, as they had to cope with an increasingly complex way of life with more and more visual signals. At the same time, the use of scent messages gradually decreased, and so primates' faces became flatter, with dry, less pointed noses.

Primates are very adaptable animals. They are found in a variety of *habitats*, from hot, humid *rainforests* to cool, dry, rocky mountains. Some monkeys even feel at home in the snows of Japan or in the high Himalayan mountains.

Like all primates, this young rhesus macaque (top) will learn the social rules from its parents. Above: a male proboscis monkey in the mangrove swamps of Borneo. The male gorilla (facing page) is the world's largest primate.

MONKEYS AND APES

2
Lemurs

Lemurs are strange, beautiful primates that are found only on Madagascar, an island in the Indian Ocean. Madagascar is the fourth largest island in the world and its wildlife is unique. For millions of years animals and plants on Madagascar have largely been isolated from those on the African mainland.

Nobody can be sure how the primitive primates of fifty million years ago got to Madagascar. Perhaps some clung to trees which were swept out to sea from the mainland, to be washed up onto Madagascar's shores. However they got there, their descendants found few other mammals with which to compete. They spread out to occupy all the different habitats on the island and evolved into the twenty-three species of lemur we know today. These range in size from the tiny lesser mouse lemur, whose head-and-body length totals four inches, to the indri, which is twenty-seven inches long from its head to the base of its tail. Sadly, though, we have just missed seeing the largest lemurs.

Less than 2,000 years ago, when people first began settling on Madagascar, they hunted the animals, chopped down the forests for their farms, and introduced cattle and goats. The island has never been the same since. At least fourteen species of lemur were made *extinct,* including a lemur that was as big as an orang-utan. As the human population continues to

With its long muzzle and moist nose, the lemur (below left and facing page) has a highly developed sense of smell. The arboreal sifaka (below) adopts this characteristic gait when it comes down to the ground.

grow, and the few remaining forests are felled, many more species of lemur are facing extinction today.

Where they are protected, lemurs still leap through the trees with grace and agility. They are mainly *arboreal* animals, and much of each day is taken up with feeding. The way in which different lemurs find food is particularly interesting.

Over millions of years, lemurs have adapted to play the same roles in the forest as other, more familiar, animals do in other countries. Like monkeys, some lemurs eat leaves and fruit. The gentle lemur specializes in feeding on bamboo, like pandas, and the mongoose lemur feeds on nectar, like a humming-bird. Perhaps the strangest lemur, if not the strangest primate, is the aye-aye, which plays the role of a woodpecker. Aye-ayes have chisel-shaped *incisors* and a special middle finger which is long and thin, with a sharp, curved claw at the end. They use this to winkle out beetle grubs from holes in rotten wood. First they use their chisel-like teeth to enlarge the hole, then they insert their finger into the enlarged hole, hook it into the grub, and pull it out to eat it.

3

Bushbabies, Lorises and Pottos

In some ways, bushbabies and lorises are rather like lemurs, but they don't live in Madagascar. The shape of their teeth, for example, has hardly changed since the first lemur-like primates of fifty million years ago. Yet whereas the true lemurs have had Madagascar to themselves for most of that time, bushbabies have had to share their habitat with higher primates. This may explain why all the members of this group are *nocturnal*. By feeding at night, they avoid competing with monkeys that might be looking for the same food sources during the day.

Their huge, round nocturnal eyes and furry face give bushbabies an irresistibly cute appearance, though the name probably stems from their sad, childlike cry. Bushbabies feed mainly on insects. They catch their prey by waiting on a branch until a moth or beetle flies past. Their ears are so sensitive that, like a bat, they can follow the sound of a flying insect until, with a lightning-fast grab, they snatch it out of the air.

A bushbaby can move around its *territory* quite rapidly, often making huge leaps to get from branch to branch. This method of travel works just as well in dense forest as it does in the more scattered trees of the *savannah*. It often disturbs insects, which the bushbaby can then capture and eat; and it also means that, by leaping, it can usually escape from any *predator*.

With their large eyes and sensitive ears, bushbabies (below and facing page top) are well adapted to a nocturnal existence. Facing page bottom: the angwantibo, a member of the loris family.

Lorises and pottos, however, have developed the opposite method of getting about and avoiding predators. They move incredibly slowly, rather like a chameleon stalking along a branch. At the slightest hint of danger, a potto or loris will freeze, and remain motionless – for hours if necessary – until the danger has gone. They also have a second line of defense, even though they have lost the ability to jump. If a potto or loris meets a large snake, or something equally frightening, it will simply let go of its branch and plummet through the dense foliage to the forest floor.

Being a slow mover is a problem for an *insectivore*, because all the tasty moths and grasshoppers can easily avoid capture. As a result, pottos and lorises feed mainly on slow-moving things, which they track down by using their sense of smell. Caterpillars with irritating hairs, foul-smelling beetles and poisonous millipedes – which would be rejected by other, faster insectivores – are all eaten by these "slow-motion" primates.

4
Marmosets and Tamarins

Facing page: a cotton-top tamarin, from the forests of Columbia, signals to others by showing off its striking tufts. Weighing around 4-6 ounces, the pigmy marmoset (above left) of Upper Amazonia is the smallest species of monkey in the world. Golden lion tamarins (above) are almost extinct in the wild, but some zoo-bred groups have been reintroduced into the forests of Brazil.

This group of fast-moving, alert little monkeys is found only in the forests of South America. All of the twenty-one species of marmosets and tamarins are small, but what they lack in size, they more than make up for in color and character. As a group, they sport an amazing array of ear-tufts, moustaches and bushy manes. These brightly colored fur patterns help them to send visual signals through the dense foliage of their forest home.

Marmosets and tamarins have an unusual family life compared to other primates. A male pairs up with a female and they stay together for life. The female usually has twins, and almost as soon as they are born the father helps to carry and look after them. Within six months another set of twins will be born, but the first set will already be feeding itself. The young reach sexual maturity at between one and two years of age, but they do not begin to breed at this point. Instead, they stay with their parents and help to carry babies. They also help by catching nourishing insects to feed the babies and their mother. New twins continue to arrive twice a year, so an extended family group might have from four to fourteen individuals, but there is only one breeding female at a time.

The diet of these miniature monkeys is very varied. They eat fruit, flowers and nectar, as well as small animals, such as lizards, frogs, snails and insects. Only marmosets, however, are adapted to make use of another, more unusual source of food. Like the aye-aye, marmosets have long, chisel-like incisor teeth, but instead of chiselling dead wood for grubs, marmosets chisel holes in living wood and then lap up the gummy sap that trickles out. Pygmy marmosets are particularly fond of this food, and will spend up to two-thirds of their feeding time eating gum. Each hole will give only a small amount of gum, so after licking for one or two minutes, they will move on to a new site and gouge out another.

Many of the different species of marmoset and tamarin have evolved in relatively small areas of forest. In the upper Amazon region, for example, each kind is separated by rivers from those in the neighboring forests. Now that these areas are being opened up for timber removal and farming, forest animals such as these are in danger of extinction.

5
New-World Monkeys

Monkeys from the forests of South and Central America – the so called "new world" – are different in many ways from those found in Africa and Asia – the "old world." If you look at their front end, you will see that new world monkeys have sideways-opening nostrils that are set wide apart on their face, whereas old world monkeys have narrow, downward-pointing nostrils that are close together. If you look at their back end, you will see that many old world monkeys have patches of hard skin on their bottoms, whereas new world monkeys all have furry bottoms.

There is also another feature at the back end of larger-bodied new world monkeys which is not found in any other primate: they have developed a fifth limb. Spider monkeys, woolly monkeys and capuchins all have *prehensile* tails. This means they can curl the tip of their tail around a branch and easily suspend their full weight from it. The tail is able to grip so well

With its prehensile tail, the spider monkey (facing page top and above) lives up to its name. Squirrel monkeys (facing page bottom) live in thickets and swamps on the edge of taller forests.

because the underside has developed an area of bare, padded skin at the end, which looks exactly like the palm of a dark hand. There is even a pattern of tiny ridges, like a fingerprint, on this sensitive area.

This "extra hand" is especially useful when a monkey is feeding on fruits that are growing at the tip of a tree branch. Clutching the thicker part of a branch with its tail and hind legs, the monkey can often reach out to the fruit with both arms. Spider monkeys are very good at this, and although they have only five "legs," not eight, their name does describe well their spider-like clambering through the branches.

Another group of large, new world monkeys includes some of the loudest animals in the world. The howler monkey does not, as the name suggests, howl like a dog; in fact its call is more like a roar. It is so loud it can be heard up to a mile away in the forest. Every morning at first light, and again whenever the troop moves on during the day, the male howler monkey calls. This is to let any nearby troops know where his family is. When howler monkey troops meet, the males run around the branches howling, and sometimes a fight may break out. It is much easier, therefore, for each leading male to avoid meeting another troop by loudly advertising his location in the forest.

6
Old-World Monkeys

The monkeys of Asia, Africa and Gibraltar are the largest and most varied group of primates, as well as the most familiar. The group includes forty-six species of baboons, guenons, and macaques, plus thirty-seven species of colobus and leaf monkeys. They live in virtually every kind of habitat there is on land - from treetops to grasslands and cliffs - but all of them walk and run on their hands and feet. Even the ones that live in trees run along the tops of branches; they never swing along underneath them like spider monkeys.

Colobus, leaf monkeys and forest guenons spend most of their time in the trees. They know the pathways along the branches of their trees like we know the streets of our home town. If one monkey makes a spectacular leap across a gap between the *canopy* of two trees, the rest of the troop will almost certainly leap across the same gap. Sometimes, their falling weight springs a long, supple branch down and it may thump the ground before swishing them back up again. Some species, such as the black and white colobus monkey of Africa, can get around so well on their treetop trails that it is rare to see them on the ground. Occasionally, though, they have to descend from the trees to cross a wide glade, or to drink from a stream. It is at such moments that tree-dwelling monkeys are most vulnerable to predators.

The main food of leaf monkeys is, as their name suggests, leaves, although fruit, flowers, buds, bark, roots and shoots may also be eaten. Leaf and colobus monkeys both have unusual

Shown relaxing in the sun, gray langurs are regarded as sacred animals by the Hindus. This species is often found around villages and temples in India.

The vervet, green or grivet monkey is the only guenon that does not inhabit forests.

stomachs because of their diet of vegetation. Like a cow's, their stomach has extra chambers, and these contain bacteria which help to digest the leaves. Perhaps the strangest looking leaf monkey is the proboscis monkey. The male's nose is so long that when he looks up to select a leaf, it flops back and smacks him between the eyes!

On safari in the African bush, the two most commonly seen primates are the vervet, or green monkey, and the local species of baboon. Vervets are typical guenons. They live in large troops that search in the grassland around the forest edge for seeds, fruit, nestlings and insects. Baboons also live in large troops, but they are much bigger. Like the vervets, they are *omnivores*. Their large size, though, means that they are not limited to catching small animals: they can hunt together to kill larger prey, such as gazelle kids.

7
Lesser Apes

The white-handed gibbon, or lar (above and facing page bottom) is found in the rainforests of Thailand, on the Malay peninsula and in northern Sumatra. The siamang (facing page top), about twice the size of most other gibbons, has an inflatable throat sac that enhances its call.

As the sun rises over the tropical *rainforests* of Southeast Asia each day, it is greeted by a beautiful sound. This dawn chorus is different from the one heard by early risers in Britain or in North America. Instead of the melodious whistling of woodland songbirds, people who live in the rainforest, from Bangladesh and Burma to Malaysia and Indonesia, wake up to the song of a mammal – the gibbon.

There are nine species of gibbons, or lesser apes. Most of them are not much bigger than a large monkey, but gibbons can easily be recognized by their extremely long arms and their lack of a tail. Gibbons are superb acrobats, perfectly adapted to a life in the treetops. They can travel at great speed through the forest canopy, swinging from arm to arm beneath the branches. This graceful form of travel is called brachiation and gibbons are the best *brachiators* in the world. When a branch is too thick for their long fingers to hook over, they walk upright on top of it, like a man, with their arms held out to balance themselves. They walk on the ground like this too, sometimes.

A male gibbon pairs up with a single female for life. The two of them then stake out a territory together by singing. Gibbon songs are elaborate duets of whoops and wails. As well as keeping the neighbors at bay, singing together also serves

to strengthen the pair's relationship. The loudest part of the duet is sung by the female, and is known as her "great call." Each species of gibbon has a characteristic call.

Usually, there is only one species of gibbon calling in any particular patch of forest. The different species are all separated from each other, with one exception, by rivers or stretches of sea. The single exception is the siamang, a large black gibbon species that shares the forests of Malaysia with the lar gibbon, and those of Sumatra with the agile gibbon.

All gibbons are primarily *frugivores*, with varying amounts of young leaves and a few insects making up their diet. They prefer ripe, pulpy fruits, and they only weigh some eleven to fourteen pounds, the males being about the same size as the females. They can swing themselves out to the end of branches. There, they hang by one hand and pluck and eat the surrounding fruits with the other.

8

Great Apes

There are three species of great ape in Africa today: gorillas, chimpanzees and bonobos (also known as pygmy chimps, though they are not much smaller than common chimps), while in the Far East, on the islands of Sumatra and Borneo, lives the fourth: the orang-utan. Perhaps, if the mysterious footprints in the Himalayan snows are ever traced to their source, a fifth may one day be described: the yeti.

In all the known great apes, males are bigger than females; an adult male gorilla or orang-utan may weigh twice as much as one of his females, but the difference is less pronounced in chimps and bonobos. The largest primate on earth – with a height of up to six feet – is the adult male gorilla, known as a silverback. He does not

Chimpanzees (left) live in large, roving communities, and an individual may meet, socialize and part with many other chimps in the course of a day. Above: a silverback mountain gorilla leads from behind, bringing up the rear in case danger should threaten.

The orang-utan – the name means "man of the woods" in Malay – spends most of its time alone, feeding in the trees.

look particularly tall, because an ape stands on all fours, on the flat of his feet and the middle knuckles of his hands. It is the width of a gorilla that is particularly impressive. A silverback can weigh more than four hundred pounds, and may have an arm span of nine feet. He leads his family group of females and their young on their daily search for food in the forest.

Gorillas are almost entirely vegetarian. They eat parts of many different plants; each mouthful, whether it be a handful of leaves, a peeled stem, bark, fruit or roots, is carefully selected and prepared before it is eaten. Some plants in their diet are found high in the forest *canopy*. Gorillas can climb very well, if cautiously, but most of their time is spent on the ground.

Chimpanzees and bonobos, on the other hand, are lighter and more agile in the trees. They eat more fruit than gorillas, and so spend more of their feeding time in trees. Chimps can brachiate quite well, but to travel any distance they always come down to the ground and "knuckle-walk" like a gorilla. Bonobos spend more time in the treetops than chimpanzees, but orang-utans are by far the most arboreal of the great apes. The orang is the largest tree-dwelling animal of all. Too heavy to brachiate, it clambers and swings its way from fruit tree to fruit tree, and only occasionally descends to the ground.

When evening falls, great apes of all four species build nests to sleep in. Every ape builds its own nest, except for infants, who snuggle down for the night in their mother's arms. Gorillas usually nest on the ground, but all the others (and even some young and female gorillas) build nests in trees, weaving branches together into a strong, comfortable platform.

9
Family Life

Monkeys and apes make very good parents. When baby primates are first born, they are helpless and depend on their mother for food, comfort and warmth. Except for the marmosets and tamarins, which usually give birth to twins, one baby at a time is the general rule among primates. And to make sure their offspring grow up safely, primate parents invest a lot of time in caring for each infant.

A newly born baby monkey or ape cannot eat any solid food. In order to survive, it must find its first meal of mother's milk. The baby makes little rooting movements with its head. Its hands automatically grip hard onto its mother's fur, but its other movements are jerky and uncoordinated. An experienced mother will gently position her baby, so that it soon locates her nipple. For the first part of its life, the baby primate never loses contact with its mother. From within the warmth and security of her hairy arms, it looks out and begins to learn about primate family life.

Primate families come in all shapes and sizes. The basic family bond – between mother and child – is nearly always the strongest. The role of the father, however, is much more variable when it comes to parenthood. At one extreme lies the orang-utan: the male orang mates with a female, then has no further contact with her or their baby for about four years; the *gestation* period lasts nine months, and it takes three years for the baby to be weaned off the breast and onto solid foods. Only then, when the female is

Surrounded by snow, Japanese macaques and their young enjoy a hot bath in geothermal springs.

receptive to mating again, does the male spend any time with her. At the other extreme is the male marmoset, which takes the baby at birth and carries it around and cares for it (except for breast-feeding!) until it is able to feed itself.

Most species of primates live in social groups consisting of a male and several females. There may be a dozen or more females, depending on the strength of the male; his task is to try to build up his harem of females, and prevent rival males from running off with any. Thus, although each female has only one infant at a time, there are usually other infants of a similar age in the group. Once an infant has reached the stage of taking its first steps – whether it is along a branch or on the ground – another, slightly older infant will probably be waiting to play with the new family member.

Above: undaunted by her large male baby, a red-fronted lemur mother climbs through the trees of southern Madagascar. Right: a female mountain gorilla and her young.

10

Play and Grooming

Young monkeys and apes are very much like human children. They are very sociable, and like to make friends with other youngsters of about the same age. As monkeys and apes cannot talk, they form friendships by playing and grooming together.

Play behavior is an important part of growing up in most mammals. First, it is very good exercise; rough-and-tumble play develops strength and coordination. Second, and perhaps more important, play has several social functions. The games that primates play are often based on adult behavior patterns. Some of these patterns are *instinctive*; others arise when young animals imitate the behavior of their parents. Because monkeys and apes are so intelligent, new games are often made-up by inventive individuals, and then copied by others. However they begin, games such as mock-fighting, mock-courtship and "pretending-to-look-after-baby" are all good practice for real activities in later life.

While they are playing, the youngsters are getting to know the strengths and weaknesses of their playmates. Each animal will thus discover his or her place in the *hierarchy* and whether a childhood playmate could be a future mate, ally, or potential rival. When a relationship does begin to develop during play, it is often reinforced later by the act of grooming.

Monkeys and apes are meticulous about grooming. With dexterous fingers and lips, they carefully pick through their own fur (self-grooming), or each other's fur (social grooming). Self-grooming is primarily concerned with keeping clean and keeping warm (because if the fur gets matted with dirt, it does not keep out the cold). Social grooming is also concerned with hygiene – a friend or relative can reach the parts that self-grooming cannot reach – but it is more than this. Primates seem to love being groomed, and this activity strengthens a family bond or friendship between individuals. A monkey will signal the beginning of a grooming session by lip-smacking, with a little movement of the head, or by staring intently at a piece of the other's fur. The other will then present a flank or belly to be groomed and lie back, often with eyes closed in apparent bliss, as particles of dirt, twigs or grass seeds, flakes of skin and dried scabs are all carefully removed. Ecto-parasites, such as lice and their eggs, are also picked off and crushed between the teeth. Social grooming may sometimes go on for one or two hours at a stretch – the silent equivalent of a long human conversation.

Ring-tailed lemurs (top and facing page bottom left) live in large troops, so there is never any shortage of playmates. Above: a chimpanzee's swaggering walk is part of his display of strength. Social grooming, among Japanese macaques (facing page top) and baboons (facing page bottom right), is how monkeys and apes make friends.

11

Life in the Forest

Primates are perfectly adapted for a life in the trees, both physically and behaviorally. In the preceding pages we have seen the range of physical adaptations; let us now look at their behavior. A day in the life of a forest primate will depend to a large extent on what it eats.

If the animal is a frugivore, it must first locate a tree bearing ripe fruit. Monkeys and apes have good color vision, which enables them to spot colored fruit and flowers from some distance. They also have accurate memories, and probably have a fair idea which fruit trees in their home range are likely to be nearing ripeness. After sunrise, therefore, the fruit-eater must travel from its sleeping tree or, in the case of great apes, its nest site, to find its breakfast. The fact that ripe fruit trees tend to be few and far between in the forest means that a fruit-eater needs a much larger home range than a leaf-eater.

The baby gorilla (facing page top) learns which foodplants to eat by picking up the remains of its mother's meal. Squirrel monkeys (facing page bottom) are able to reach food that larger monkeys cannot reach. Above: precariously perched gray langurs.

Leaf-eaters can usually just reach out an arm to begin feeding in the morning. In fact, gorillas quite often begin the day with breakfast in bed by reaching out and pulling up the nearest foodplant, then peeling it and eating it while still in their night nest. As a result of this abundance of food to hand, leaf-eaters do not usually travel far each day, and they tend to *forage* in larger groups.

If a monkey is concentrating on collecting and eating leaf-buds or bamboo shoots, it cannot always be on the lookout for danger. By foraging for food together, a troop of monkeys benefits from the larger number of eyes looking out for predators. This strategy would not, however, work so well for a fruit-eater. A troop of large fruit-eating primates would soon finish off the ripe fruit at a single tree, and would then have to spend time searching for another food supply. This is especially true of apes, with their enormous appetites. Orang-utans are largely solitary, gibbons live in pairs, and although chimps live in large, loose communities, they split up to forage. Gorillas, on the other hand, live and forage in groups of up to twenty-five or thirty animals.

After the morning feeding time, and before the afternoon session, forest primates usually have a rest period. The troop will settle down to relax in the sun, monkeys up in the branches and gorillas and chimps in a secluded glade. Adults doze and groom, while the younger animals chase and play. They do not post a lookout because in a forest you hear danger approaching long before you see it through the leaves.

12

Life on the Plains

Vervet monkeys (above) have a range of alarm calls that warn of different dangers. Facing page top: Chacma baboons retreat to the security of an old tree during the heat of the day. Facing page bottom: a baboon drinks at a hotel swimming pool.

The species of monkey which have moved out onto the plains include baboons, macaques, vervets and the patas monkey. The patas is an extraordinary monkey because it has evolved long legs for running across grassland. However, it does this between clumps of acacia trees where most of its food is found, and its diet is similar to that of forest guenons, to which it is related. Baboons have adapted to living on the plains by including a lot of grass in their diet. Grass is very difficult to digest, and so baboons supplement this with almost anything else they can lay their hands on. They are among the most successful of hunting primates, and will work together to flush out, chase and kill hares and young gazelles. If there is enough food in their home range, baboons may live in troops of up to 150 animals. They have sharp eyes and, at the first sign of danger, the lookout gives a warning "waa-uu" bark and everyone else is instantly on the alert, ready to run for cover or up a tree.

Vervet monkeys use different alarm calls to warn of different predators. When a vervet sees an eagle, it makes several sharp grunts and the other troop members all climb down from the tops of trees to take cover. If one sees a leopard it makes a different call; this produces the opposite effect – everyone rushes up the nearest tree and out onto the thinnest branches. And a high pitched chattering call makes everyone stand up on two legs and look around in the grass; it means a snake has been seen.

Baboons and macaques are great opportunists, and will make use of any new

supply of food, including that from human rubbish heaps or even hotel balconies. A new feeding behavior will soon be copied by other troop members, and then passed on to new generations as a part of the culture of that troop. This was seen very clearly during a study of Japanese macaques. The scientists put out a supply of sweet potatoes on a beach so they could watch the monkeys in the open. A young female named Imo soon began washing the sand off her sweet potatoes, first in fresh water, then in salt water, which apparently tasted better. When wheat was scattered on the beach, Imo invented a new technique. She threw a handful of wheat and sand into the sea; the sand sank and the floating grains could then be scooped off the surface of the water. Both habits were copied and became a part of that troop's culture.

The behavior of savannah-dwelling primates is of particular interest to anthropologists. It gives us an insight into the origins of the behavior of the most successful primate on earth – the one which now threatens all other primates – *Homo sapiens.*

13

Danger to Monkeys and Apes

Destruction of their natural habitats, the trade in live animals for pets, and the slaughter of monkeys and apes for their skins means that many primates face the very real possibility of extinction.

Primates are hunted by many different predators, and an attack can come from any direction. Monkey-eating eagles, for example, with especially enlarged talons for crushing monkey skulls, soar over the forests of Africa and the Philippines; leopards, and other cats, crouch ready to leap from trees and rocks; pythons lie in wait on branches, coiled ready to strike, and crocodiles lurk, like lethal logs, at the water's edge. And from the forest floor, far below the canopy, unseen human hunters fire blow-pipe darts, or arrows, or shotgun pellets.

Such dangers pose a threat to individual monkeys and apes, but they do not normally wipe out a whole population, or threaten the survival of a complete species. And yet, many species of primate are today faced with extinction. The problem is the same one that faces most other endangered species: there are too many humans.

Human beings threaten their fellow primates in a variety of different ways. In South America, hydro-electric dams flood vast areas of forest, and rich rainforest is being felled to be replaced by cattle ranches; small populations of monkeys, marmosets and tamarins are particularly vulnerable to the destruction of their sole patch of forest.

In Africa, monkeys and apes are shot for food and sold in markets as "bush-meat"; monkeys with attractive skins are shot for tourist souvenirs; and live baby primates are sold as pets, as well as to zoos and biomedical research laboratories. Smaller primates can be trapped or netted when adult, but to capture a baby chimp or gorilla, the parents and any other defensive group member must first be shot. Many infants also die in this cruel and wasteful trade, and so reputable zoos will no longer buy primates caught in the wild.

In Southeast Asia, forests are dwindling as timber companies fell trees faster than they can grow replacements, and baby gibbons and orang-utans are still sold as playthings for wealthy people, despite laws prohibiting the trade. There are tough conservation laws in many countries which have wild primates, but they seldom offer total protection because the laws are difficult to enforce.

In the long term, the single most important threat to primate survival is the loss of their natural habitat. As human numbers continue to rise, the pressure to build on or cultivate every patch of "unused" land mounts. Numbers of most monkeys and apes have steadily declined during the past century. Only if we human primates can reverse these trends, will the non-human primates of this world have a future.

Glossary

ARBOREAL – living in the trees, seldom descending to the ground.

BRACHIATOR – an animal that travels by hand over hand locomotion, hanging beneath branches.

CANOPY – the layer of leaves at the top of trees.

EXTINCT – when the last remaining member of a species dies.

FORAGE – the act of searching for food.

FOSSIL – the mineralized remains of an ancient animal or plant found in rock.

FRUGIVORE – an animal that eats mainly fruit.

GESTATION – the period of development of a baby mammal, from the fertilization of an egg to the birth.

HABITAT – the natural home of any plant or animal.

HIERARCHY – the order of importance between individuals in a social group, sometimes known as the "pecking order."

INCISORS – the front teeth, used for cutting.

INSECTIVORE – an animal that eats mainly insects and other terrestrial invertebrates (land animals without backbones).

INSTINCTIVE – behaviour that is controlled by genes, and so is inherited by each new generation.

NOCTURNAL – active at night.

OMNIVORE – an animal that eats almost anything.

PREDATOR – an animal that kills and eats other animals.

PREHENSILE – able to grasp.

RAINFOREST – a forest that depends upon a high rainfall to grow (most, but not all, of the world's rainforest lies between the Tropic of Cancer and the Tropic of Capricorn).

SAVANNAH – flat, open grassland with occasional trees.

SPECIES – a particular kind of animal or plant that breeds true.

TERRITORY – an area which is marked out and/or defended for exclusive use by an animal or group of animals.

Picture Credits

ANIMALS ANIMALS 11 *top left*; G.I. Bernard 28; Mike Birkhead 18, 24 *bottom*; Scott Camazine 31; Robert P. Comport 5; Margot Conte/ANIMALS ANIMALS 16; Phil Devries 4 *bottom*; Carol Farnetti 8, 9 *top*;. Michael Fogden 12 *bottom*, 13, 26 *bottom*; David C. Fritts 21; Mickey Gibson/ANIMALS ANIMALS 20; Esao Hashimoto 25 *top*; Peter Lack 1; Bertram G. Murray Jr./ANIMALS ANIMALS 17 *top*; Patti Murray/ANIMALS ANIMALS 29 *top*; Stan Osolinski 2, 15, 25 *bottom right*; Partridge Productions 9 *bottom*; Mark Pidgeon 6, 7, 22, 24 *top*, 25 *bottom left*; Andrew Plumptre 19, 23, 26 *top*; Ian Redmond 29 *bottom*, 30; Frank Roberts/ANIMALS ANIMALS 11 *top right*; Leonard Lee Rue III/ANIMALS ANIMALS 12 *top*; Alastair Shay 17 *bottom*; Babs and Bert Wells 10; Belinda Wright 4 *top*, 14, 27.